Biological Classification of Molecules

Organic Molecules and Carbon Molecules

Martin H. Gremlich

Institute for Human Biology

Editorial

Copyright © 2020 Martin H. Gremlich

All rights reserved. No part of this publication may be reproduced and/or transmitted by any means and in any form, graphic, electronic, manual, or mechanical, including photocopying, recording, taping or by any information storage retrieval system without prior written consent of the author.

Gremlich, Martin H. / Biological Classification of Molecules – Organic Molecules and Carbon Molecules

Published by Institute for Human Biology
www.human-biology.org
Design and concept Martin H. Gremlich

Library of the Institute for Human Biology
Catalog Nr. K11400 001

ISBN 978-1-7771390-2-5

To Aristotle
(384–322 BC)

Aristotle taught us that knowledge alone is useless. We need to critically analyze acquired knowledge and make sense of it by means of consequent logical thought in order to create individual understanding.

Biological Classification of Molecules
Organic Molecules and Carbon Molecules

Content

Introduction	1
The Scientific Purpose of Classifying	5
Analysing the two Classifications	9
Historic Classification (Organic)	9
Recent Classification10 (Carbon)	14
Dissolving the Irrational Academic Dispute	18
Logical Conclusion	22
Relevance of The Two Classifications	26
Relevance of Carbon Molecules	28
Relevance of Organic Molecules	31
Summary: Organic and Carbon Molecules	38
Final Note	46
Appendix	51
History of the Original Classification (Organic)	51
History of the Recent Classification (Carbon)	54

Introduction

Science – to differentiate science from numerous other academic activities – has only one purpose: To create understanding.

Today, however, there are two seemingly conflicting academic classifications of all molecules, which both claim to differentiate between "organic" and "inorganic" molecules. The present dispute in academia does *anything but* creating understanding – it creates confusion and misunderstanding.

The two classifications of "organic" molecules differ fundamentally.

The original (historic) functional classification "organic"

An "organic" molecule is a molecule, which cannot exist naturally as result of chemical reactions within the biosphere of the planet, unless the molecule is (or has been) metabolized by a living organism. All other molecules are "inorganic" molecules.

The recent elemental classification

An "organic" molecule is a molecule, which contains one or several atoms of the element C (carbon) – with some exceptions. All other molecules are "inorganic" molecules.

Some academic proponents of the recent classification declare the original classification to be "wrong" or "outdated" for some not logically obvious reason. In addition, they declare that their recent classification has exceptions. Some molecules are declared as being "inorganic" *despite* containing the element C.

Functional Biology is the encompassing logical natural science of Life. It studies living organisms in context (including human organisms) and intends to create understanding by logically building on the known laws of Physics and Chemistry.

Like everything else, living organisms *consist* of molecules – as can be observed.

In order to study life scientifically (including *human* life), we need to *consider* the most fundamental framework of our matter-space-time universe – the physical magnitude dimension of atoms and molecules, the domain of molecular Chemistry

Consequently, *both* classifications may principally be useful to create understanding.

However, we cannot simply *ignore* this irrational academic dispute; we must logically clarify and *dissolve* the dispute for good.

In order to achieve this, we need to proceed systematically.

First, we think about the general scientific purpose of any classification whatsoever.

Subsequently, we analyze the two classifications of molecules in detail and clarify them, so that we can discover the actual cause of the present intolerable academic dispute and logically dissolve it.

Finally, we assess both classifications with respect to their usefulness and relevance as scientific tools in specific situations with the goal to create understanding in the fundamental life science Functional Biology.

The Scientific Purpose of Classifying any Array of Objects

Scientific classifications differ fundamentally from academic conventions (non-rational declarations or theorems).

Scientific classifications are never "wrong" or "outdated". Classifications are simple logical tools, which may be either useful or irrelevant in certain situations.

As tools, classifications can only serve to order any otherwise unordered array of objects (e.g. all molecules) into two sub-arrays.

To create a classification tool simply requires defining a distinguishing criterion (any criterion). A criterion is a characteristic (any characteristic), which preferably only some (but not all) objects in a given array of object have in common.

A criterion (characteristic) allows sorting all objects into two smaller groups (sub-arrays) – one sub-array containing all objects, which do fulfill the criterion and the other sub-array containing all objects, which don't.

As can be observed, both classifications of "organic" molecules discussed here provide such a criterion – one criterion is "stemming from an organism" the other is "containing the element C (carbon)".

Logically, both classifications are equally valid ordering tools. Hence, both classifications can principally serve scientifically to sort all molecules into two sub-arrays.

None of the two classifications can logically replace or invalidate the other. Scientifically, *different* classifications can be applied to *the same* array of objects without ever disqualifying each other – they simply define different sub-arrays.

The number of different but logically valid classifications possible for the same array of objects (e.g. all molecules) is principally unlimited.

Each different classification simply defines two unique sub-arrays (without ever having the potential to scientifically disqualify any other classification).

There is a fundamental logical classifying condition, however. There must be *strictly no exception* to the classifying criterion.

The moment we declare any exception to the classifying criterion, we create *a mess* instead of order. There are now some objects in one sub-array *because* they fulfill the criterion and some objects (the exceptions) in the *other* sub-array *despite* fulfilling the criterion.

The other sub-array no longer contains only objects, which do not fulfill the criterion; it now also contains some, which do (the exceptions).

Since this is no longer logical, we have to *learn by heart* (or consult an index) to determine to which sub-array each object is academically declared to belong.

Any classification whatsoever, which allows any exception, becomes *logically invalid*. It is no longer a scientific ordering tool and is reduced to an academic insider convention.

Scientific Condition

In the domain of the logical natural life science Functional Biology, *both* existent classifications may be *useful* in some situations. However, in order to have them *both* available, we must *strictly forbid* any exception to the recent classification with its classifying criterion "contains the element C".

All carbon-containing molecules must be *in the same* sub-array of molecules, while exclusively (only) all molecules, which do not contain the element C, must be in the other.

Only under this logical condition do we have *both* classifications available scientifically. Otherwise, we'd be left with only one logically valid classification of organic molecules – the *original* classification (criterion organic = stemming from an organism).

Analysing The Two Classifications

Under the logical condition that there are no exceptions – In order to determine in what situation which of the two classifications may serve best to create understanding, or whether it even may be useful to apply both classifications simultaneously – we may analyse them both.

The Original Functional Classification (Organic)

The classifying criterion:

All molecules exclusively metabolized by an organism vs. all others.

A molecule is classified as "organic", if under the biospheric conditions on the planet the particular molecule must be – or must have been at some point in time – generated by a living organism.

The molecule cannot be observed (or cannot be chemically explained) to form otherwise within the biosphere. Some *organic* examples are e.g. O_2, ATP, RNA, DNA, etc.

All other molecules are called "inorganic" molecules.

Important to consider

The sub-array "inorganic" includes also all molecules, which can be *both*, generated by a living organism (as result of its metabolism) as well as form and exist in the biosphere otherwise – without any living organisms having generated them.

Such *inorganic* molecules are e.g. H_2O, CO_2, CH_4, etc.

The *criterion* of this classification of molecules is *functional*. The criterion is a molecule's *origin*, *generator* (environment or living organism), or its *provenance*.

This criterion lies in the physical magnitude dimension of living organisms. The classifying criterion requires consideration of the molecular biospheric condition on our planet (a biological expertise).

The historic criterion "organic" cannot be determined chemically. For this classification it is completely irrelevant whether a molecule contains the element C or not.

The classification can be applied to every known molecule. Each molecule can principally be conclusively allocated to one specific of the two sub-arrays.

The classification does not define any exceptions.

Consequently, the historic classification is of course *scientifically valid* (today as much as ever).

Figure 1

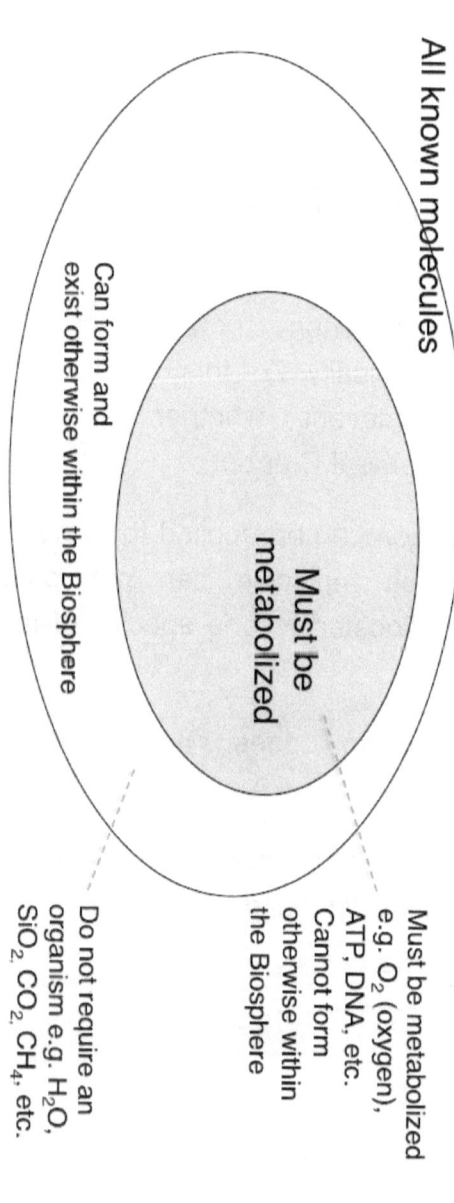

The word "organic" literally means "stemming from an organism".

Hence, the *name* Organic Molecules for this functional classification is *rational*.

The name includes the classifying criterion, which makes the classification as such interdisciplinary comprehensible.

See Appendix
History of the Original Classification

The Recent Elemental Classification (Carbon)

The classifying criterion:

All molecules containing the element C (carbon) vs. all others.

The classifying criterion lies in the magnitude dimension of atoms and molecules. It cannot be determined functionally.

The criterion differentiates molecules by the presence of one or several atoms of the element C in its structure or the absence of any element C.

For this classification, the origin, generator, or provenance of a molecule is completely irrelevant.

Under the logical condition that there are *no exceptions*, this classification forms two clearly identified sub-arrays of all molecules.

Figure 2

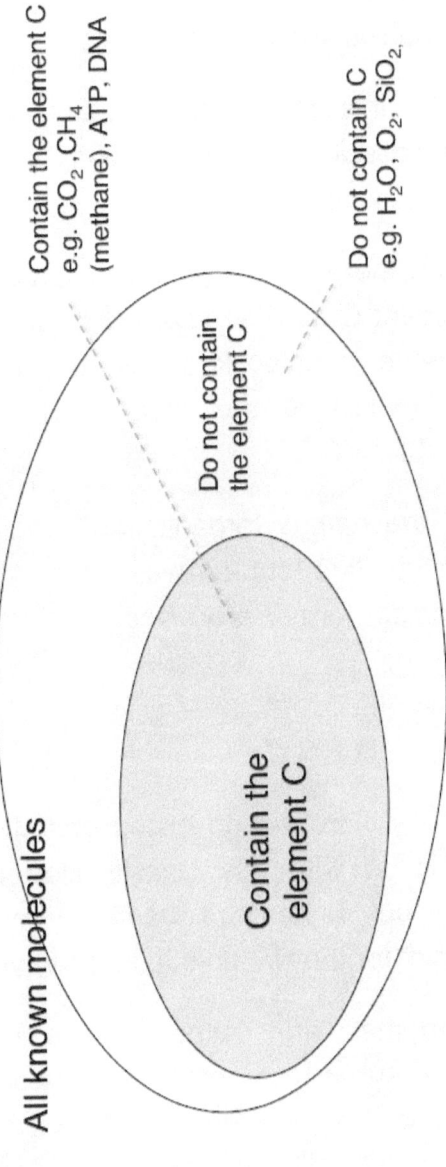

The classification "contains the element C" allows allocating each molecule conclusively to one of the two sub-arrays it forms.

Logically, this recent classification is perfectly valid as a scientific classification, as long as it allows no exceptions.

However, there is no logical connection between the element C as a structural part of a molecule and whether a molecule stems from an organism or not (which is the meaning of the word "organic").

Hence, the *names* "organic" and "inorganic" for the two unique sub-arrays, which the recent classification forms, are of course *irrational* and misleading.

As long as there are no exceptions allowed, the logical issue with this classification is not its criterion but rather *the names* "organic" and "inorganic" for the two unique sub-arrays it forms.

These names are neither unique names for molecular sub-arrays nor does the word organic ("stemming from an organism") have anything logical to do with whether a molecule contains the element C or not.

This renders the recent classification of course ambiguous and *incomprehensible* across scientific disciplines.

See Appendix:
History of the Recent Classification

Dissolving the Irrational Academic Dispute

The academic dispute has arisen with the recent academic insider decision of molecular chemists to usurp the names "organic" and "inorganic" for their recent and unique carbon-based classification of molecules, instead of simply creating unique and adequate names for the two sub-arrays it forms.

Obviously, the two classifications order all molecules into *different* sub-arrays.

Neutrally analyzed, there is no detectable scientific conflict.

There is simply an irrational academic dispute about *ownership* of the word "organic" (which means "stemming from an organism).

A single name "organic" for two entirely different but logically equally valid classifications of molecules must of course create an insurmountable ambiguity.

This renders *both* classifications practically useless, since scientists can identify none of them conclusively as long as there is just one name for both.

Such an ambiguity is principally *intolerable* in any logical natural science.

However, the ambiguity can easily be dissolved. We simply assign a *unique name* to each of the two classifications.

Assigning a unique name to each classification does not affect any of them logically in any way, nor does it affect their respective classifying criterion or the unique sub-arrays each of them forms.

In logical natural science, it is always advisable to assign a (unique) *name* to a classification tool, which indicates the *classifying criterion*.

A classification name, which contains the criterion enhances ease of understanding (which is the *purpose* of science in general).

It also facilitates academic *interdisciplinary* communication and helps avoiding specialized disciplines locking themselves up in academic crystal palaces.

Consequently, in order to create clarity and understanding, we simply assign the following unique name to each of the two classifications – at least valid for the domain of the encompassing natural life science Functional Biology:

Organic Molecules (the historic classification)

are all molecules, which must have been generated by a living organism to exist within the biosphere. All other molecules are *inorganic* molecules (the name "organic" indicates the *criterion* "stemming from an organism").

Carbon Molecules (the recent classification)

are all molecules, which contain the element C (no exceptions). All other molecules are *non-carbon* molecules (the word "carbon" indicates the *criterion* "contains the element C").

Note

It is of course theoretically possible, that two different classifications could form two *identical* sub arrays.

In this case, it would be acceptable (even though scientifically not practical) to use a single name for both classifications, since both classifications would order all molecules identically.

However, as can be observed, this is most definitely *not the case* with the two classifications Organic Molecules and Carbon Molecules analysed here.

Each of the two classifications obviously orders all molecules into two *different* sub-arrays.

Hence, an identical (single) name for both classifications (and their differing sub-arrays) is scientifically *principally inacceptable*.

Logical Conclusion

There is no detectable logical conflict between the two classifications.

The two classifications are simply two different logical ordering tools, which create two different sub-arrays of all molecules.

Under the strict logical condition that there are no exceptions, *both* classifications – Organic Molecules and Carbon Molecules – are scientifically *equally valid* as classification tools to order all molecules.

The two classifications may be applied *simultaneously*. If simultaneously applied, the two classifications form *four* distinct logical sub-arrays.

Figure 3

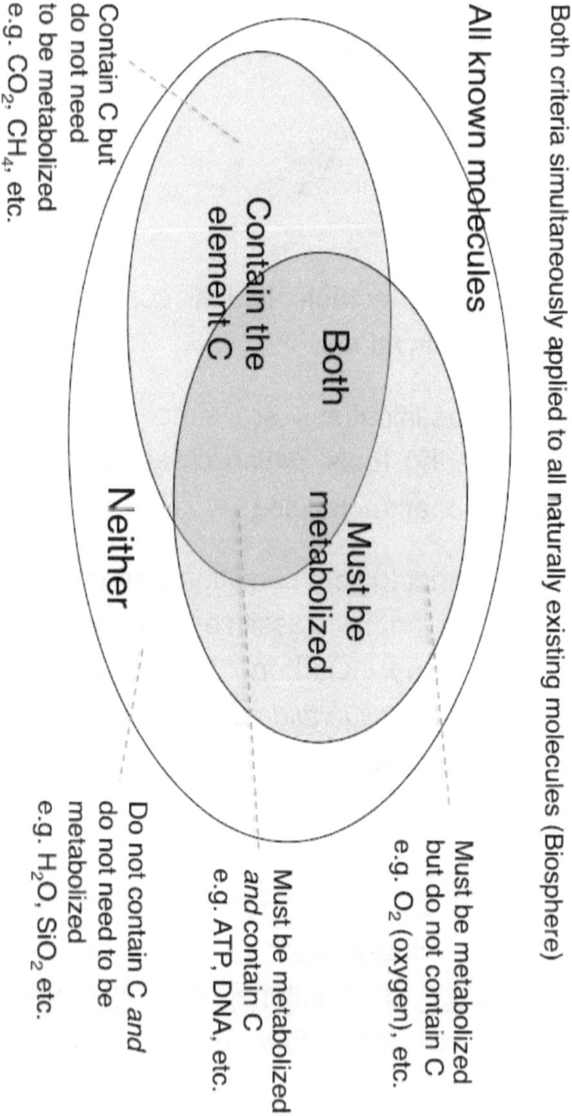

Both criteria simultaneously applied to all naturally existing molecules (Biosphere)

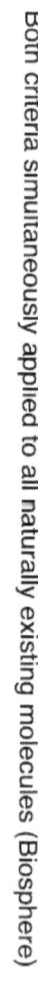

All known molecules

Contain the element C

Both

Must be metabolized

Neither

Contain C but do not need to be metabolized
e.g. CO_2, CH_4, etc.

Must be metabolized but do not contain C
e.g. O_2 (oxygen), etc.

Must be metabolized *and* contain C
e.g. ATP, DNA, etc.

Do not contain C *and* do not need to be metabolized
e.g. H_2O, SiO_2 etc.

The four logical sub-arrays are:

Organic non-carbon molecules
> Molecules, which must have been metabolized by an organism but do not contain the element C (e.g. O_2, etc.)

Organic carbon molecules
> Molecules, which fulfill *both* criteria. They must be metabolized by a living organism *and* contain the element C (e.g. ATP, DNA, etc.).

Inorganic carbon molecules
> Molecules, which contain the element C but exist also in the biosphere without mandatorily having been metabolized by an organism (e.g. CO_2, CH_4, etc.)

Inorganic non-carbon molecules
> Molecules, which fulfill *neither* criterion. They are neither mandatorily metabolized by an organism nor do they contain the element C (e.g. H_2O, N_2, SiO_2, etc.)

Both classifications applied simultaneously allow a more refined scientific ordering of all molecules than each of them alone, which is of course principally *to the benefit* of all logical natural sciences.

Scientifically, the classifications Organic Molecules and Carbon Molecules are not "either or"; they most definitely are "and".

Both classifications are equally valuable ordering tools in a logical scientific toolbox.

None of them can *replace* the other. Each of them can serve to create understanding in a specific situation. Each of them is tailored to a specific use.

Relevance of The Two Classifications for Functional Biology

The question in logical natural science is of course never the *validity* of a logically valid classification.

The question is simply the *usefulness* or *relevance* of any specific classifying tool when attempting to order objects (in this case all molecules).

The relevance of a classifying tool depends (among other considerations) also on the physical magnitude dimension of the classifying criterion, the physical magnitude dimension we study.

Important to consider

The criteria of the two classifications discussed here lie in entirely different physical magnitude dimensions of our matter-space-time universe.

In other words, we preferably select the classifying tool, which is best suited or relevant in the magnitude dimension we are thinking, studying, or researching in order to create scientific understanding.

Relevance of The Classification Carbon Molecules

The classifying criterion is "the molecule contains the element C". This criterion lies in the physical magnitude dimension of atoms and molecules:

This magnitude dimension of atoms and molecules is of course the core foundation of our matter-space-time universe.

It is the magnitude dimension, where all atomic bonds and chemical reactions happen. In this dimension, atoms and individual molecules interact with each other. This is the magnitude dimension of Chemistry.

The chemical magnitude dimension is an intellectual comfort zone. In this dimension things happen *automatically*, *reliably*, and *predictably*. It is these reliable automatisms, which have allowed us to discover the so-far known Laws of Chemistry.

Today, some expert astronomers assume the element C (carbon) to be the 4th-most numerous element in our Milky Way galaxy, which definitely would make carbon a *most significant* element in the chemical dimension

Apart from this estimation, in the magnitude dimension of Chemistry and all its academic sub-domains (e.g. molecular Physiology etc.), the unique chemical characteristics of the element C are indeed most relevant.

Atoms of the element C have the unique capacity to form particularly strong – even cyclic – atomic bonds with other C molecules. Such solid carbon-structures apparently form atomic "skeletons" of most of the larger known molecules.

As long as we think, study, and research in the automatic magnitude dimension of atoms and molecules and their individual chemical interactions, the classification Carbon Molecules may be very useful and relevant.

However, in the magnitude dimension of Chemistry (and all its academic sub-domains) we must principally assume (or even observe) a molecule to being present either physically or theoretically. Without an atom or molecule at least assumed to being present, we simply have no object to study.

Consequently, since a molecule must be assumed to *being present*, in the magnitude dimension of Chemistry (and its sub-domains), the actual origin, provenance, or availability of any specific molecule (the criterion of the original classification Organic Molecules) is basically *irrelevant*.

However, even though the classification Organic Molecules is simply not useful or relevant in the automatic magnitude dimension of atoms and molecules, it logically remains of course a perfectly valid scientific classification tool, which is by no means "wrong" or "outdated" today.

On the other hand, in the magnitude dimension of Chemistry it is simply impossible to differentiate scientifically between inorganic objects and living organisms. As long as we think, study, or research in the chemical magnitude dimension, we remain restricted to chemical automatisms, since there aren't any *living* nor any *non-living* molecules – as far as we know today.

Relevance of The Classification Organic Molecules

The magnitude dimension of living organisms – the magnitude dimension of the historic criterion "organic" vs. "inorganic" – is no longer the dimension of individual molecules automatically interacting with each other.

It is a *functional* magnitude dimension of specific complex *molecular mega structures* existing in a molecular and energetic context (Biosphere).

There is an enormous magnitude and complexity gap between the automatic dimension of individual molecules and the molecular mega structures of living organisms (the smallest of which consist of estimated *at least billions* of specific individual molecules of an overwhelming diversity).

In this magnitude dimension, we differentiate between two fundamental types of molecular mega structures – *inorganic* (non-living) objects and *living organisms* (the scientific domain of Biology).

The scientific differentiation between inorganic objects and living organisms requires a typical valid logical classification with a respective criterion.

However, in the magnitude dimension of molecular mega structures, the scientific classifying criterion lies no longer in the chemical dimension of atoms and molecules, it is a *functional* classifying criterion.

We differentiate living organisms scientifically from inorganic objects by their different functional mode of physical existence as molecular mega structures – an active molecular existence (living organisms) vs. a passive molecular existence (inorganic objects).

Inorganic objects of any size (from a mini-object of only two molecules to molecular structures the size of galaxies and beyond) form and exist *automatically*, *passively*, *reliably*, and *predictably* as directly dictated by the laws of Physics and Chemistry.

Hence, the laws of Physics and Chemistry valid for the *molecular* magnitude dimension of atoms and molecules can be extrapolated and applied to all *inorganic* molecular mega structures of any shape, size, and form.

Existence of *inorganic* (non-living) objects is principally passive, automatic, and predictable from a molecular level upward.

Living organisms on the other hand – as can be observed and chemically confirmed – *cannot* form and exist automatically, passively, reliably, or predictably directly dictated by the laws of Physics and Chemistry within the biosphere.

The physical existence of living organisms obviously requires their *active, autonomous* molecular *self-construction* and self-maintenance as molecular mega structures.

No living organism can possibly just "pop up" within the biosphere as the result of reactions of available inorganic molecules with each other. If they could, they would have to automatically pop up – time and again (enforced by the automatic laws of Chemistry).

Living organisms must self-build many (organic) molecules they observably consist of. Living organisms obviously cannot simply "harvest" these molecules from their environment, since these molecules obviously cannot form there (e.g. DNA),

Physical existence of living organisms –already in the molecular dimension – is principally *active* and *autonomous* (non-automatic) *from a molecular level upward*.

Consequently, chemical automatisms *cannot* simply be logically extrapolated into the magnitude dimension of living molecular mega structures (living organisms) – even though doing so is apparently a most common rather simplistic academic practice (e.g. Physiology, Pharmacology, Medicine, etc.).

Chemical automatisms are not the *reason* for existence of living molecular mega structure. These automatisms are *the means* every living organism must *actively and selectively engage* under its environmental conditions (biosphere) in an autonomously self-regulated way to self-build, self-energize, and self-maintain its own molecular mega structure (called anatomy).

This also includes the mandatory self-generation of certain specifically required but otherwise unavailable molecules, which are classified as *organic molecules* (whether they contain the element C or not).

As soon as we think, study, or research in the infinitely greater magnitude dimension of *living organisms* it is no longer relevant, whether a particular required molecule contains the element C or not.

In the functional magnitude dimension of living organisms (the scientific expertise of Functional Biology), the *origin, availability, presence or absence* of each specifically required molecule – both *within* the organism and *in its imminent environment* – becomes crucially significant.

Consequently, the original (historic) classification Organic Molecules becomes *relevant* and extremely *useful* to create scientific understanding.

In this magnitude dimension, however, it is *irrelevant* whether a functionally required specific molecule contains the element C or not (e.g. H_2O does not contain the element C).

Hence, the classification Carbon Molecules is neither relevant nor useful in this dimension (yet, with no exceptions allowed, it is a perfectly *valid* classification).

Note

The *idea* "Organic Molecules" (molecules mandatorily *stemming* from an organism) has directly led to the discovery of a fundamentally vital *continuous activity* of living organisms – the activity of metabolizing *specifically required* molecules (with or without an element C).

The activity of metabolising molecules must be autonomously (non-automatically) self-regulated by an organism to produce the *right* molecules, in *the* right amount, in *the* right sequence, and at the *right* time (whether they contain the element C or not).

This cannot be explained with chemical automatisms alone (based on Carbon Molecules). It must be explained *functionally* (based on the concept of Organic Molecules – specific molecules that cannot be "harvested" from the inorganic environment but must be self-generated by an organism).

What can be explained chemically, though, is the fact that some of these obviously existing molecules simply couldn't form and exist under the biospheric conditions on the planet, unless they are (or have been) metabolized by a living organism (e.g. O_2, ATP, gigantic DNA molecules, etc.),

This of course confirms the relevance of the historic classification Organic Molecules *chemically* in the magnitude dimension of living organisms.

Summary: Organic and Carbon Molecules

There is no detectable logical conflict between the historic and the recent classification of molecules. Logically, both, the historic and the recent classification are valid, as long as neither allows any exceptions.

However, there is an intolerable irrational academic dispute about the ownership of the names "organic" and "inorganic".

Both scientific classification tools are principally only useful, if each classification can be *identified conclusively*.

This of course first and foremost requires a *unique name* for each classification – preferably indicating the unique criterion of the classification.

At least in the domain of the encompassing natural life science Functional Biology, we must *clearly distinguish* between Organic Molecules (stemming from an organism), a *functional* classifying criterion, and Carbon Molecules (containing the element C), an *elemental* (atomic) classifying criterion.

This distinction is scientifically necessary because the two classifications obviously form *different* (non-identical) sub-arrays of all so far discovered biospheric molecules.

The recent (renamed) classification Carbon Molecules

The classifying criterion is "contains the element C (carbon)".

This criterion lies in the magnitude dimension of atoms and molecules (the magnitude domain of Chemistry). In this automatic and predictable magnitude dimension, the element C plays a unique and significant role. Classifying molecules by presence or absence of the element C in a molecule may be useful and relevant.

However, there are no "living" or "non-living" molecules as far as we know today.

In the automatic magnitude domain of Chemistry (and its sub-domains e.g. molecular Physiology) we principally *cannot* differentiate between a living organism and an inorganic object.

Any particular molecule "looks" chemically exactly the same, whether an organism has metabolized it or not.

Hence, in the chemical magnitude dimension of Chemistry the historic classification Organic Molecules is *not applicable* (yet scientifically perfectly valid).

The historic classification Organic Molecules

The criterion "organic" (stemming from an organism – the origin or provenance of a molecule) lies in the most significantly greater physical magnitude dimension of living molecular mega structures in their molecular and energetic context (the domain of Biology).

The smallest organisms discovered so far consist of at least *billions* of specific molecules of an enormous variety forming a specific molecular structure (an anatomy).

It becomes *relevant*, whether a specifically required molecule is available – whether it simply can be "harvested" from the environment or must be *self-generated* – metabolized – to become available.

Hence, the *availability*, *origin*, or *provenance* of a particular molecule becomes a crucial classifying criterion in the magnitude dimension of living organisms.

Consequently, the classification Organic Molecules is useful and relevant in the magnitude dimension of living organism to create scientific understanding.

The classification Carbon Molecules is irrelevant and not useful in this dimension, since many non-carbon molecules are equally significant in this dimension (e.g. the molecule H_2O).

As soon as we can *distinguish* the two classifications by a unique name, *both* valid classifications of molecules may be applied simultaneously.

This allows a more refined scientific ordering of all molecules, which may be scientifically useful and relevant in some situations.

Important to consider

The recent classification Carbon Molecules principally cannot *replace* the historic classification Organic molecules scientifically – simply because not *every* organic molecule contains the element C and not *every* carbon containing molecule must be generated by an organism to exist within the biosphere.

The classification Carbon Molecules orders all molecules *differently* than the classification Organic Molecules.

Further, the *classifying criterion* of each of the either classification lies in a fundamentally different physical magnitude dimension.

The question is not the *validity* of any of the two valid classifications (they are both *equally* valid, as long as there are no exceptions).

The question is rather their *usefulness* and *relevance* as a classifying tool in the magnitude dimension we are thinking, studying, or researching.

As long as we concern ourselves with specific chemical reactions of individual molecules with each other (no matter whether within or outside living organisms), the classification tool Carbon Molecules is useful to understand chemical interactions.

The classification Organic Molecules, on the other hand, is perfectly valid but cannot be applied in the magnitude dimension of Chemistry (the classifying criterion "stemming from an organism" cannot be identified on a molecular level).

Living organisms (living molecular mega structures) obviously require an enormous variety of specific molecules (both, carbon and non-carbon) in specific quantities at specific locations in a specific arrangement at specific times.

As soon as we think, study, or research in the magnitude dimension of living organisms (Biology), the classifying tool Carbon Molecules is still perfectly valid but no longer useful or relevant, since in this dimension non-carbon molecules (e.g. H_2O) are just as significant as carbon molecules.

In the magnitude dimension of *living* molecular mega structures (organisms) the crucial factor is the *availability* of each required molecule (no matter whether carbon or non-carbon).

It becomes important to consider whether an organism can simply "harvest" a required molecule from its environment (as is) or whether the organism has to *self-generate* (metabolize) the molecule to have it available at all.

In the physical magnitude dimension of living organisms of any size, shape, or form the classification Carbon Molecules is not useful (it is valid but irrelevant).

On the other hand, the classification Organic Molecules is fundamentally important to create scientific understanding in this magnitude dimension.

Final Note

Obviously – in praxis – it is intellectually substantially more demanding to classify a particular molecule as organic or inorganic than it is to classify the same molecule as a carbon or a non-carbon molecule.

We cannot classify a molecule as organic or inorganic simply by analysing its atomic structure (e.g. in a lab). We need to go far beyond the intellectual comfort zone of chemical automatisms.

A frequent misunderstanding of the classification Organic Molecules is that organic molecules could only exist *within* living organisms, which is in fact true for the vast majority of organic molecules..

However, some organic molecules are chemically pretty stable. Some of them may have existed already for billions of years *outside* a living organism (e.g. O_2, $CaCO_3$, etc.). In fact, calcium carbonate ($CaCO_3$) is so stable that it has become a significant component of sedimentary rocks over time (e.g. limestone, the metamorphic rock marble, etc.).

Rocks are obviously *inorganic* objects. Yet, to determine whether they consist of *organic* and/or *inorganic molecules,* we need to consider paleo-geology and paleo-climatology – the *history* of our planet and its biosphere

Otherwise, we cannot determine whether a molecule *must* have been generated by an organism to exist (the criterion "organic") or could have formed by simple chemical reactions within the biosphere of that time.

Crude oil is another example of an *inorganic* substance, which consists of an enormous variety of *organic* molecules or decay molecules thereof. Crude oil *cannot* directly form by simple chemical reactions of inorganic molecules on our planet *without* the participation of living organisms. If it could, it would have to form time and again – enforced by chemical automatisms.

Today, it is scientifically indisputable that there are some molecules existing naturally, which cannot be explained chemically to form as simple reactions of environmentally available "raw material" molecules within the biosphere (according to the known laws of Physics and Chemistry).

What can be observed, however, is that living organisms obviously do generate such otherwise inexistent molecules (e.g. enzymes, proteins, DNA, etc.).

Living organisms obviously not only generate these molecules. As also can be observed, they *consist* of them and *operate* with them in a self-regulated (non-automatic) specific and purposeful manner.

Consequently, the moment we leave the fundamental scientific comfort zone of the physical micro-dimension of atoms and molecules (the automatic dimension of Chemistry) and start to think about, study, and research the molecular macro dimension of *living organisms in context* (Functional Biology), we cannot simply extrapolate chemical automatisms into this physical dimension to create understanding.

What scientifically differentiates all living organisms most obviously from all inorganic objects already in their molecular dimension is their sovereigns capacity to actively and autonomously generate some molecules, which otherwise are unavailable for the organism in its imminent molecular environment within the biosphere (some of them containing the element C and others without).

In the physical magnitude dimension of living organisms (the competence of Biology), not only the classification Organic Molecules but also the original *functional idea* and the entire *original concept* behind it become crucial to create any scientific understanding.

The by all means significant *word* "organic" scientifically means "generated by a living organism" (nothing else).

As can be observed, an organic origin also applies to some molecules (all Organic

Molecules), which obviously exist naturally but cannot form in any other way within the biosphere than by being generated by a living organism according to the known laws of Physics and Chemistry.

Organic Molecules are *defined* (and consequently classified) as those observably existing molecules, which *must* be generated by living organisms. Otherwise they *cannot* form within the physical, chemical, and energetic biospheric conditions on the planet.

A scientific understanding of the *functional implications* of the observable existence of Organic Molecules is crucial to understand the *scientific* (logical) definition of Life on this planet (including e.g. *human* life).

— — — — — — — —

© 2020 Martin H. Gremlich
Institute for Human Biology
Module K11400 001
www.human-biology.org

Appendix

The History of the Original Classification (Organic)

Up to towards the end of the 19th century, Chemistry was basically experimental and restricted to empirical observations of "substances" and their interactions and reactions with each other.

Only a few elements had been theoretically identified, the existence of molecules could only be theorized.

By that time, it had long been obvious, that certain "substances" require a living organism to create them – e.g. blood and other body fluids or muscle tissue, bones, teeth, etc. Obviously such substances do not form outside an organism under the environmental conditions of the biosphere. This observation is of course as undisputed today, as it has been then.

Toward the end of the 19th century, microscope technology had become sufficiently potent to discover "substances" (structures) within cells, among others also structures called "chromosomes"

Today chromosomes are chemically identified as gigantic DNA molecule structures, which by the way, today is both, an organic as well as a carbon molecule.

With the development of technology at the beginning of the 20th century and fuelled by a wars and evolving oil and gasoline-burning engines (crude oil and natural gas are *organic* substances) an enormous amount of new chemical "substances" became available as by-products of crude oil refining.

A new specialized natural Science evolved – molecular Chemistry.

This rapidly led to the discovery of more elements (atoms) and elemental-structures called molecules. It soon became clear that a particularpure "substance" consists of an enormous amount of *identical* molecules, which ultimately dictate a substance's physical and chemical characteristics.

More and more different molecules were identified, among which more and more types of molecules that must have been generated by living organisms, irrespective whether they contain the element C or not.

One of the most prominent molecules, which must be generated by living organisms but do not contain any C atom is the organic gas molecule O_2 (oxygen).

This insight led to the discovery of the Great Oxygenation Events in the history of our planet and a timeline of the evolution and diversification of life on it.

Logically consequent, the identical molecules, which form a pure *organic* substance, became scientifically classified as Organic Molecules, the molecules a specific pure organic substance consists of, the molecules, which must have been generated by an organism.

Eventually, in the mid 20th century, this led to the discovery of a unique achievement of living cells called "metabolism".

Metabolism is a specific activity of living organisms *only*. It means the activity of autonomously generating specific molecules, which an organism cannot simply harvest from it environment because they cannot form there.

Non-living (inorganic) objects such as rocks, bodies of water, or air obviously never ever actively metabolize any molecules.

Metabolism does not disqualify the historic classification "organic molecules". In the contrary, it most definitely *logically confirms* the validity and *scientific usefulness* of the historic classification Organic Molecules today.

The History of the Recent Classification (Carbon)

In the mid 1950s, the periodic table of elements as we know it today had principally been established with the ground-braking integration of electronegativity values of each element (Pauling, Allen, Muliken, Alfred-Lochow, etc.).

Since then, molecular chemists have been able to calculate atomic bond strengths between atoms and reliably predict chemical reactions between individual molecules (including e.g. calculating the energy gain or consumption of a chemical reaction).

It started to become obvious that the element C does indeed play a crucial role in the physical magnitude dimension of molecules.

The element C can form particularly strong bonds with four other elements, also with other C elements. This allows the element C to form particularly stable skeletal components of most larger molecules (consisting of many atoms).

It became evident and useful to pay special attention to the element C, and to classify molecules by whether they contain it or not, in order to create chemical understanding.

As more *large* historically organic molecules were discovered particularly *within* living organisms, molecular chemists started to connect the element C directly with "life" – living organisms.

In the 1960s, space exploration started to become technically possible. This instantly launched a search for extraterrestrial life forms (today intensely ongoing).

It is assumed, that discovery of larger extraterrestrial molecules with a carbon skeleton could at least indicate molecular *precursors* (or residues) of molecules, which were potentially generated by extraterrestrial life forms, hence, may be of *organic* origin.

Around the mid-20th century, molecular chemists also started to generate *artificial* carbon molecules experimentally in labs, molecules, which are neither naturally forming in the biosphere nor metabolized by living organisms.

This launched (among other) the explosive development of today's gigantic pharmacological industry and subsequent medical applications (drugs, medication, vaccines, etc.), which indicated yet another assumed (simplistic) chemical connection of carbon-containing molecules with living organisms.

Consequently, in academic insider circles of molecular Chemistry, the element C gradually became symbolically the "life-element" – the "organic" element.

Simultaneously, since the differentiation between living organisms and inorganic objects *cannot* be made chemically (in the dimension of molecules) and the origin of a particular molecule *cannot* be determined chemically, the historic classification drifted out of focus in circles of specialized molecular chemists.

They started to declare the historic classification Organic Molecules as "wrong" and "outdated" (confounding "irrelevant for their specialized academic domain" with logically "invalid").

They simply usurped the in their eyes no longer useful name "organic" for their recent carbon-based classification of molecules – not for any apparent logical reason but probably to honour their classifying element C as the symbolic "life-element".

Again for no apparent logical reason, they also started to *declare exceptions* to their carbon-based classification. Some carbon containing molecules became declared as inorganic.

Unfortunately, this presently leaves molecular chemists and molecular physiologists de facto with no scientific classification of organic molecules at all.

While they have declared the original and valid classification Organic Molecules as being "wrong" and "outdated", while their own recent carbon-based classification Carbon Molecules is a logically invalid classification due to its declared exceptions.

Their recent denominator "organic molecules" for molecules containing the element C (with exceptions) remains a non-rational academic insider convention with a non-unique and unrelated name.

This *must* of course create academic confusion, not only interdisciplinary but particularly also among molecular chemists themselves.

A unique name for their recent chemical classification of molecules with the criterion "contains the element C" and strictly *no exceptions* would instantly create clarity – to the benefit of all scientific domains including molecular Chemistry and its sub-domains.

- - - - - - -

About the Author

Martin H. Gremlich studied natural and earth sciences at the University of Zurich and the Federal Institute of Technology, Switzerland.

After his career as a pilot and human factors expert for Swissair he founded the Institute for Human Biology in Canada of that he is currently the executive Head of Science and Education.

His goal is to create an encompassing *scientific* understanding of the functional biology of being human so that it can be reliably, practically, and proactively applied to improve individual and public health and wellbeing.

www.ingramcontent.com/pod-product-compliance
Lightning Source LLC
Chambersburg PA
CBHW031545210526
45464CB00003B/1150